What Is PHYSICS?

Rebecca Woodbury, Ph.D., M.Ed.

Gravitas Publications Inc.

What Is
Physics?

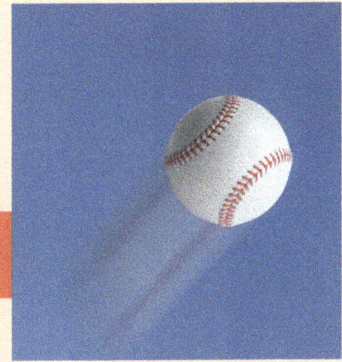

Illustrations: Janet Moneymaker

What Is Physics?
ISBN 978-1-950415-18-2

Published by Gravitas Publications Inc.
Imprint: Real Science-4-Kids
www.gravitaspublications.com
www.realscience4kids.com

RS4K

Photo Credits: Cover & Title Page: By Feng, AdobeStock; Above, By creativesunday AdobeStock; P. 3. By Marzanna Syncerz, AdobeStock; P. 8. By Beaunitta V W/peopleimages.com, AdobeStock; P. 17. Bird, By Steve Byland, AdobeStock; Popcorn, By Meg Boulden on Unsplash; Baseball, By creativesunday AdobeStock; Bowling ball, By Alec, AdobeStock; Marble, By Dan Kosmayer, AdobeStock; P. 19: 1. By David Lloyd, AdobeStock; 2. By The_Believer, AdobeStock; 3. By The_Believer, AdobeStock; 4. By NickiAnimations from Pixabay; 5. By Matthew, AdobeStock; 6. JRJfin, AdobeStock; P. 21. By Feng, AdobeStock

What happens...

...when you throw a ball?

Does the ball...

Up!

...go up?

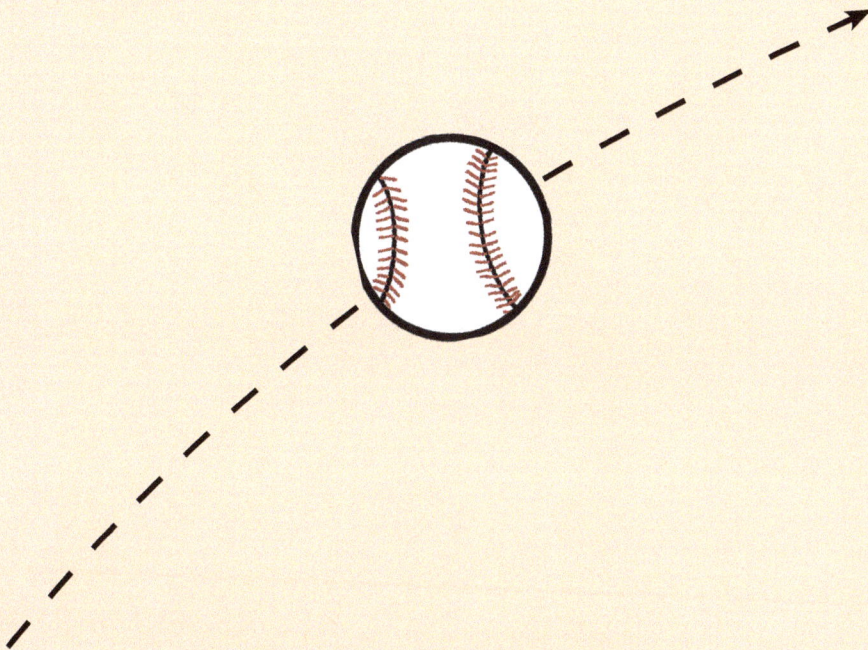

Does the ball...

What do you think?

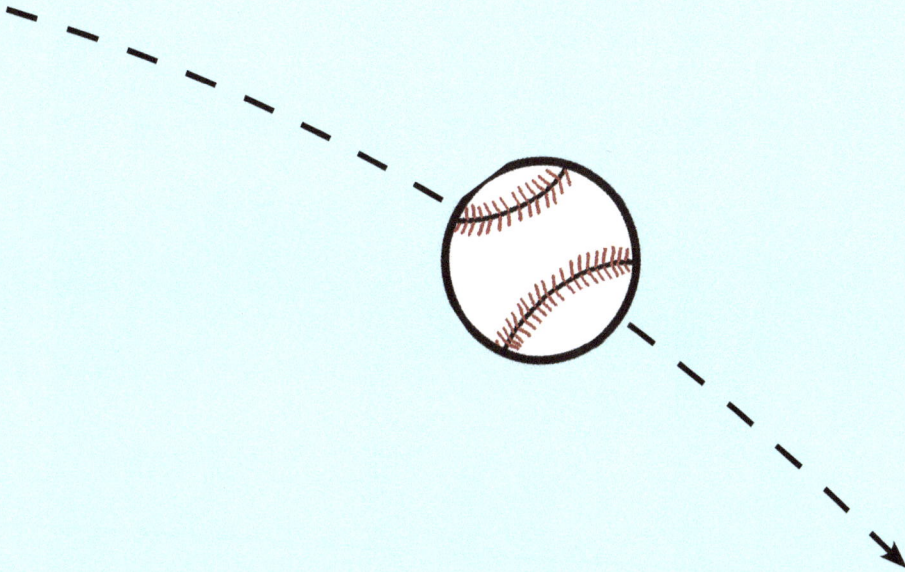

...come back down?

How far...

...can you throw a ball?

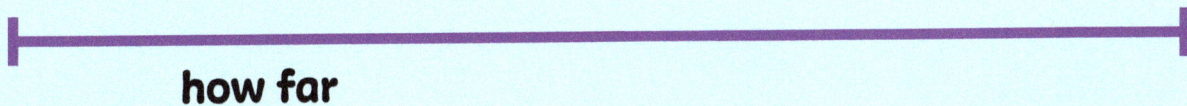

how far

How high can you
throw a ball?

how

Is it easier to throw

a ping pong ball...

...or a bowling ball?

All of these questions
are questions about
physics.

how high

how fast

how far

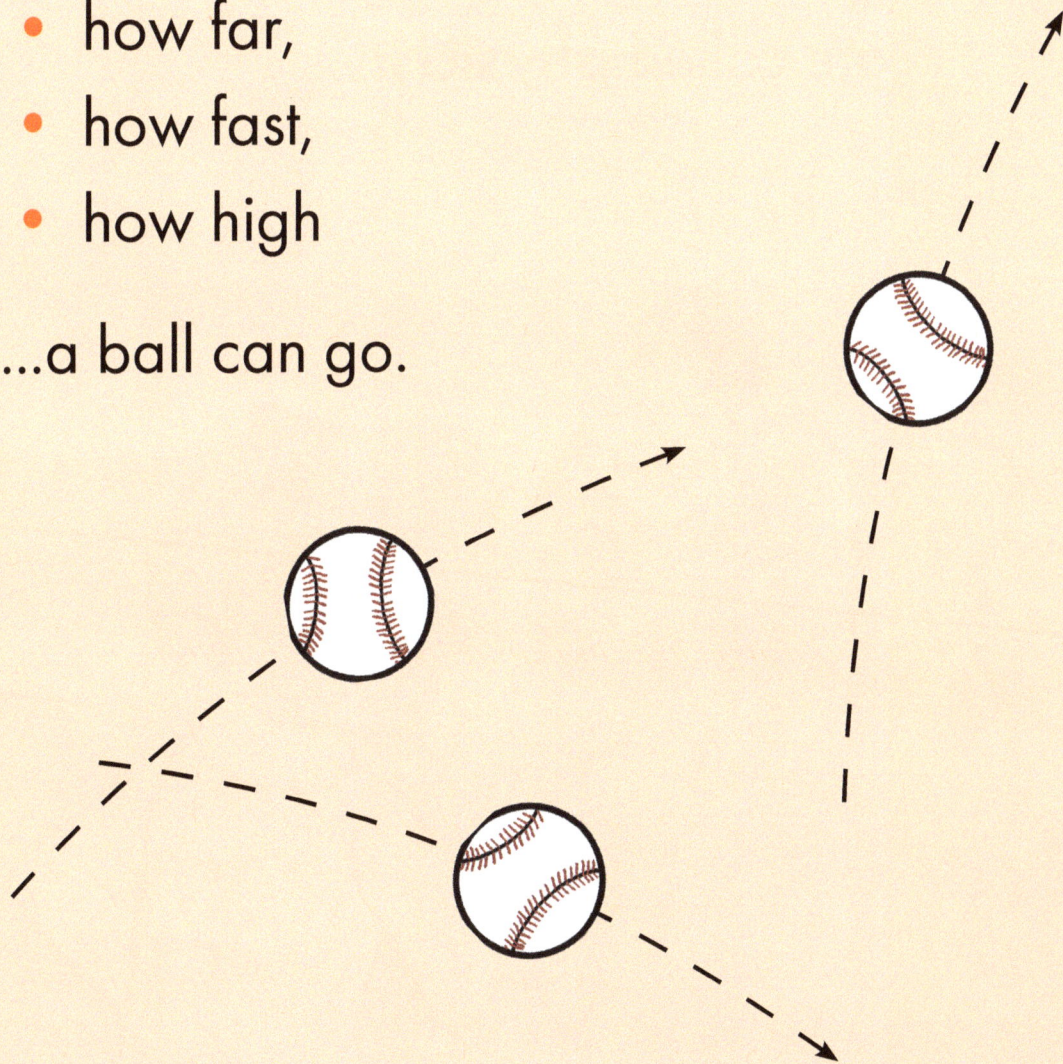

Physics is a science that asks...

- how far,
- how fast,
- how high

...a ball can go.

Scientists who study physics
are called **physicists**.

Physicists study how fast kids run.

Physicists ask lots of questions about the world.

How do birds fly?

Why does popcorn pop?

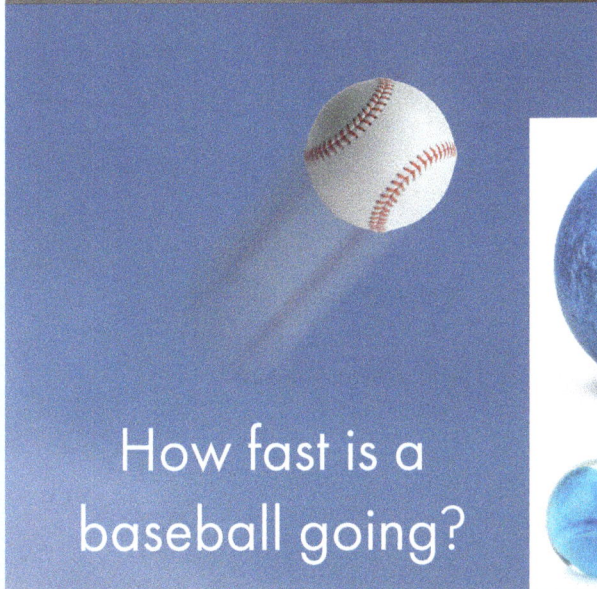
How fast is a baseball going?

How heavy is a bowling ball?

How heavy is a marble?

Physicists use many tools to help answer these questions.

Which one makes cheese?

All of them?

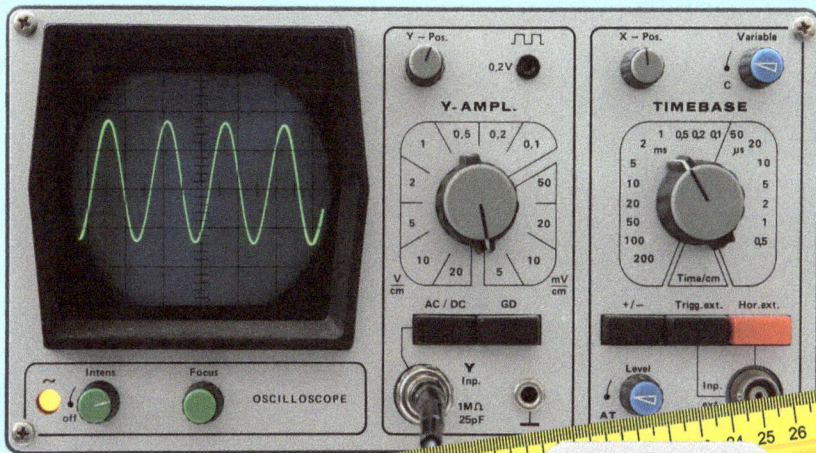

1

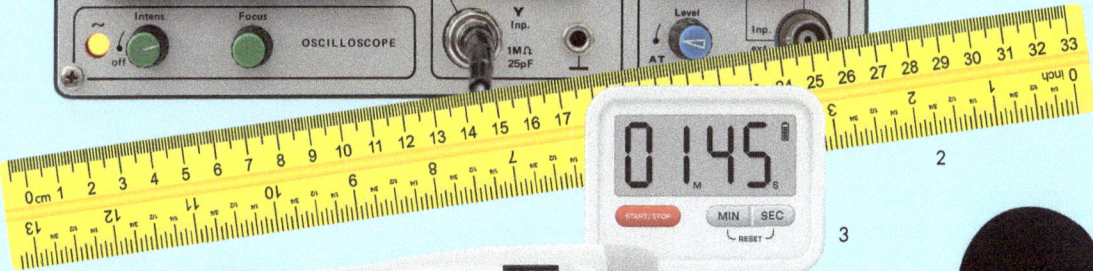

2

3

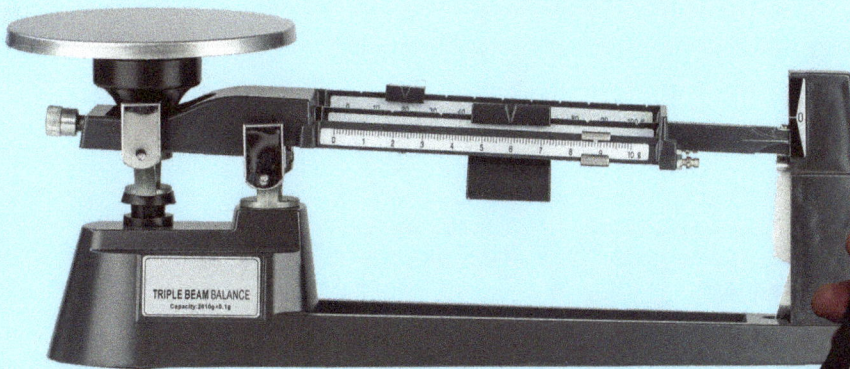

4

5

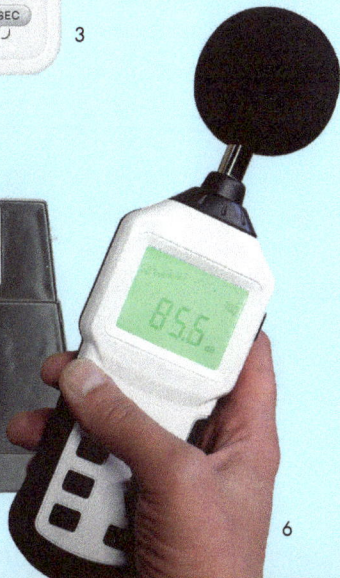

6

By using tools physicists can...

- Tell how fast a plane flies,

- Know how a ball bounces,

- Study all the colors of a rainbow.

How to say science words

physics (FI-ziks)

physicist (FI-zuh-sist)

science (SIY-uhns)

study (STUH-dee)

question (KWES-chuhn)